By Adam Montierth

Illustrated by Adam Devaney

Library of Congress Control Number: 2011922243

ISBN 978-0-9827986-5-2
Printed in the United States of America.
CPSIA facility code : BP 310351

This book is dedicated to my wife Monica.
Thank you for helping me believe that dreams can come true
and thank you for being one of them.

– Adam

Scientists say, "the composition of the Bumblebee

makes it IMPOSSIBLE for him to fly."
- but he DOES!
And with all of
his busy chores,
he doesn't worry why!

The PLATYPUS is a weird mammal,

With a BILL to match his LEGS.

I guess someone forgot to tell him,

that mammals DO NOT LAY EGGS!

Running 71 miles per hour

gives the CHEETAH a tired paw

But being above most speed limits,

It can also anger the law!

The RHINOCEROS BEETLE is very strong,

Lifting 850 times its own weight!

To us, it's like lifting A BUS!

(So we really can't relate.)

There are still things in life,

That astound the science of man.

SO when someone says
"You can't"
You just say,
"I can!"

For nothing in life

is impossible,

As long as you set your MIND to it.

All you really have to do,

Is get off your
rump and

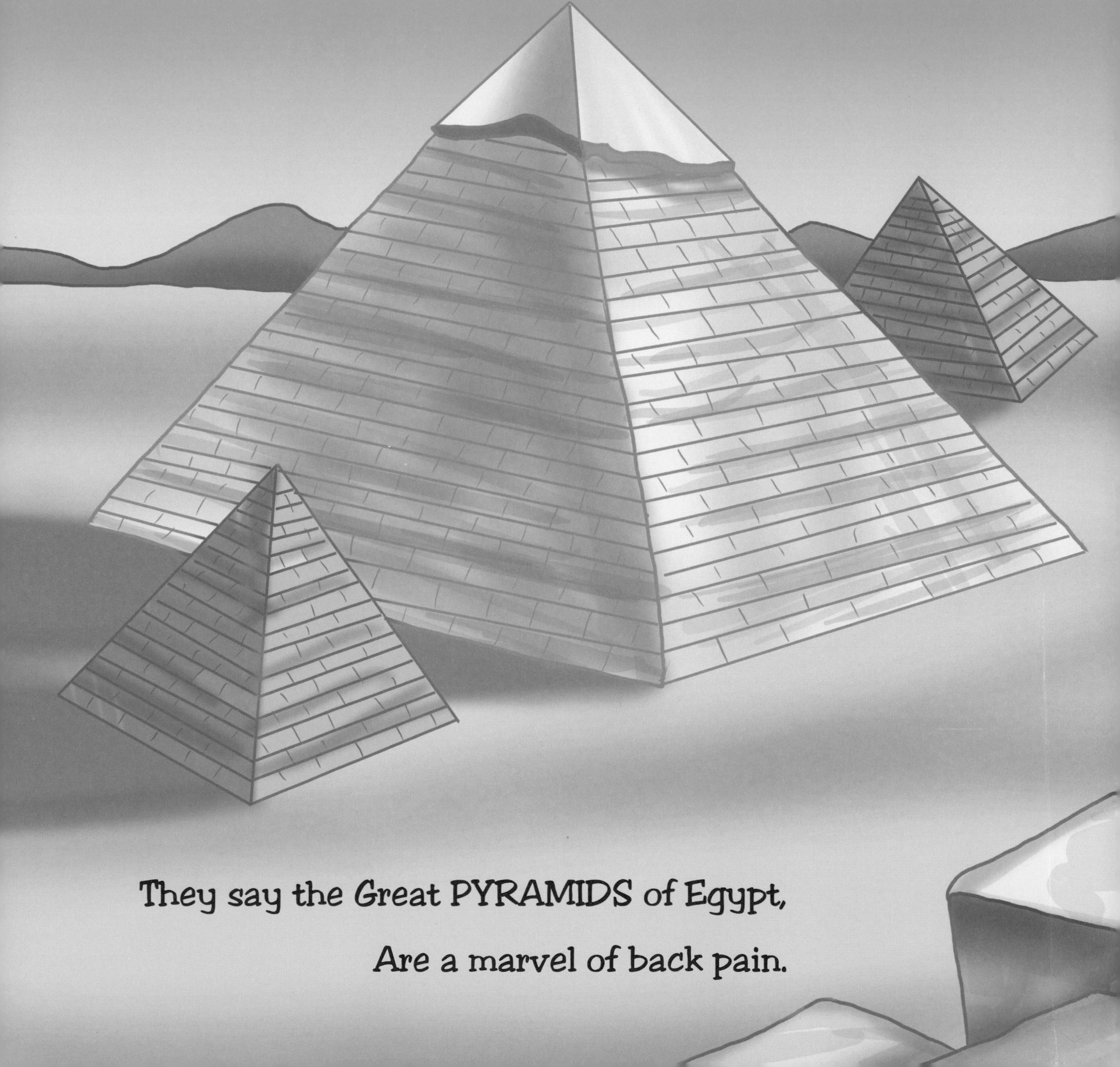

They say the Great PYRAMIDS of Egypt,

Are a marvel of back pain.

How did they move those
LARGE stones,
Without the use of a CRANE?

Being the FIRST to sail around the world,

Ferdinand Magellan knew where it was at.

In those days, they said it couldn't be done,

For most thought the world was FLAT!

The Wright Brothers proved man CAN fly

With the first controlled human flight.

To think it all got started,

With a bicycle and a kite.

In 1969, many experts said,

"Man will **NEVER** walk on the moon!"

But Neil Armstrong and Buzz Aldrin,

Soon made them change their tune!

Don't limit what you CAN DO,
When others say it can't be done.
USA
USA

Open your mind

and **dream!**

Let yourself become **Number One!**

LONG JUMP

Just remember –

a man without a dream,

Is like a bird with a broken wing.

No one recalls the cricket that doesn't sing,

Nor the man that doesn't DREAM!

Many Scientists and Engineers have said that it should be impossible for the Bumble Bee to fly. This is based solely on body size and weight compared to the size and shape of their wings. But no one ever told the bees however, so they continue to fly.

The Platypus is one of the only Mammals that lay eggs instead of giving birth to live young. When they first discovered the Platypus, it didn't originally fall into the category of "Mammals", due to its unique characteristics.

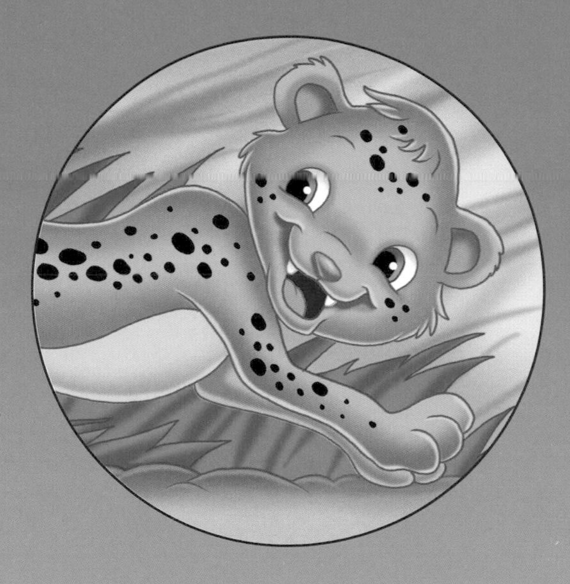

Cheetahs are the fastest land animal in the world. They can run up to 71 miles per hour, but can't maintain that speed for long. They are definitely built for speed, even without sneakers.

Rhinoceros Beetles are considered the strongest creatures on the planet, in comparison to their own size and weight. They can lift over 850 times their own weight. To put that into perspective, it's like a man lifting 65 tons. That's over ten elephants.

The Great Pyramids of Giza in Egypt are the oldest man made structures still intact. They were built some time around 2560 B.C. and they think it took around 20 years to complete. They're still trying to figure out how they moved all of the giant stones in place.

Ferdinand Magellan was a Portuguese Explorer who was in service to the Spanish Crown. In 1519 A.D., he was the first person to lead an expedition to circumnavigate the earth. He also discovered the Strait of Magellan.

In their bicycle shop, brothers Orville and Wilbur Wright dreamed of flying. With imagination, hard work, and determination on December 17th of 1903 at Kitty Hawk, North Carolina, their dreams came true with the first controlled human flight.

Man has always looked up into the night sky and dreamed about what it would be like to walk on the moon. On July 20th 1969, Neil Armstrong and Buzz Aldrin became the first men to ever step foot on the moon. Their footsteps remain there to this day, as a testament that man can do the impossible, if only they dream!